Cop

APOLC

230 Landmark Drive
Montgomery, AL 36117-2752

For information on other A.P. books call:
(800) 234-8558, or visit our Web site:
ApologeticsPress.org/catalog

All rights reserved. No part of this book may be
reproduced in any manner whatsoever without the
written permission of the publisher.

Layout by Tommy Hatfield

ISBN-13: 978-1-60063-121-4
Library of Congress: 2018938854

Printed in China

NOTE: The title for this "Learn to Read" book
was suggested by AP donor Kenny McDaniel.

Dedication
To my granddaughter – Violet

Hares, Mares, and Bears

by Dave Miller, Ph.D.

A.P. "Learn to Read" Series

God made hares.

Hares are like rabbits—but larger.

They have long ears and run fast.

They like to
eat plants.

A group of hares is a "drove."

A baby hare is called a leveret.

God made hares.

God made mares.

A mare is a female horse over three years old.

The baby of a mare is a foal.

Mares like to run.

Do you like to run?

The daddy horse is called a stallion.

God made mares.

God made bears.

Bears have large bodies.

They have shaggy hair and short tails.

Though big, bears can run, climb, and swim.

Most bears live in a den—like a cave.

They hibernate in winter.

God made hares, mares, and bears.

God made them all.

God made them all on day six.

God is great!

The "Learn to Read" Series: A Word to Parents

Rationale: To provide books for children (ages 3-6) from Christian homes for the purpose of assisting them in **learning to read** while simultaneously introducing them to the **Creator** and His **creation.**

Difficulty Level

The following listing provides a breakdown of the number and length of words in *Hares, Mares, and Bears* (not counting simple plurals and duplicates):

Total Number of Words: 64

1—One letter word
a

6—Two letter words
to, of, is, do, in, on

16—Three letter words
God, are, but, ear, and, run, eat, old, the, you, big, can, den, all, day, six

20—Four letter words
hare, made, like, they, have, long, fast, mare, over, year, baby, foal, bear, hair, tail, swim, most, live, cave, them

10—Five letter words
plant, group, drove, horse, three, daddy, large, short, climb, great

8—Six letter words
rabbit, larger, female, called, winter, shaggy, though, bodies

1—Seven letter word
leveret

1—Eight letter word
stallion

1—Nine letter word
hibernate

Drawings by Violet Dubcak, age 8

The A.P. Readers

LEVEL 1 "Learn to Read"

1. Dogs, Frogs, and Hogs
2. Bats, Cats, and Rats
3. Birds, Bugs, and Bees
4. Fish, Flies, and Fleas
5. Goose, Moose, and Mongoose
6. Ducks, Bucks, and Woodchucks
7. Snails, Quails, and Whales
8. Sharks, Larks, and Aardvarks
9. Loons, Coons, and Baboons
10. Hares, Mares, and Bears

LEVEL 2 "Early Reader"

1. God Made the World
2. God Made Dinosaurs
3. God Made Animals
4. God Made Insects
5. God Made Plants
6. God Made Fish
7. God Made You
8. God Made Birds
9. God Made Puppies
10. God Made Hair

LEVEL 3 "Advanced Reader"

1. Amazing Tamable Animals
2. Amazing Tails
3. Amazing Dinosaurs
4. The Amazing Human Body
5. Amazing Migrating Animals
6. Amazing Copies of God's Design
7. Amazing Teeth
8. Amazing Skin
9. Amazing Tongues
10. Amazing Beauty

**We continue to expand the number of titles in each
series. Be sure to check our Web site for our newest books.**
ApologeticsPress.org

READY FOR THE NEXT A.P. READER?
Try Level 2: the *Early Reader* Series

**Scan code to see a list
of all the books in the
Early Reader series**

TEACH ME AT AN EARLY AGE.
LET ME LEARN ABOUT GOD AND HIS CREATION.

A.P. Reader Series
Only $2.00 ea.

LEVEL 1 — Learn to Read — 50-250 words
New readers who sound out words and sentences
"Learn to Read"

LEVEL 2 — Early Reader — 200-750 words
Readers who are increasingly confident but still need help
"Early Reader"

LEVEL 3 — Advanced Reader — 700-1750 words
Reading fun subjects for inspiration and information
"Advanced Reader"

Bats, Cats, and Rats
Written by Dave Miller, Ph.D.
The A.P. Readers "Learn to Read" Series

God Made Reptiles
by Eric Lyons

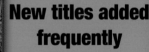

New titles added frequently

Amazing Dinosaurs
ADVANCED READER SERIES

To order call
(800) 234-8558
or visit
ApologeticsPress.org/webstore